MOTHER EARTH'S BEAUTY: TYPES OF LANDFORMS AROUND US (FOR EARLY LEARNERS)

SPEEDY PUBLISHING

Speedy Publishing LLC
40 E. Main St. #1156
Newark, DE 19711
www.speedypublishing.com

Copyright 2015

Rights reserved. No part of this book may be reproduced or used in any way or form or by any means whether electronic or mechanical, this means that you cannot record or photocopy any material ideas or tips that are provided in this book

LANDFORMS

A **landform** is a natural feature of the Earth's surface. Landforms together make up a given terrain, and their arrangement on the landscape or the study of same is known as **topography**.

VOLCANO

A **volcano** is a rupture on the crust of a planetary-mass object, such as Earth, that allows hot lava, volcanic ash, and gases to escape from a magma chamber below the surface.

Which of the two pictures is a Volcano? Put a check mark on the circle.

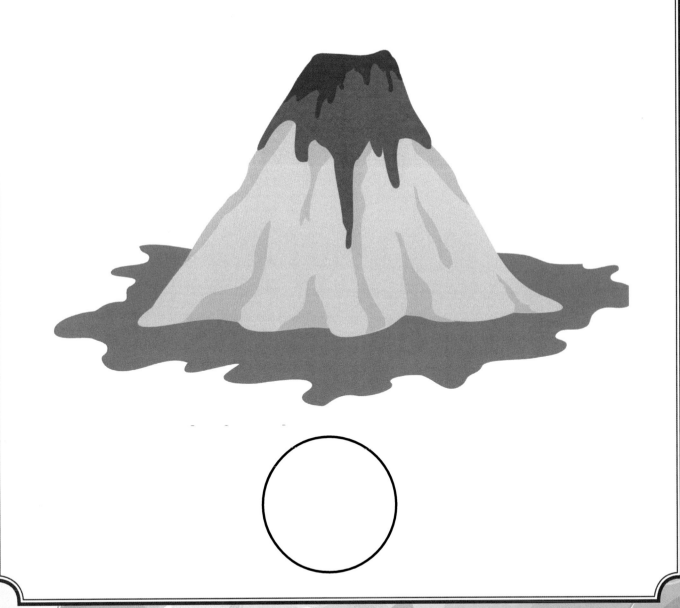

HiLLS

A **hill** is a landform that extends above the surrounding terrain. It often has a distinct summit, although in areas with scarp/dip topography a hill may refer to a particular section of flat terrain without a massive summit.

Which of the two pictures is a Hill? Put a check mark on the circle.

iSLAND

An **island** is any piece of sub-continental land that is surrounded by water. Very small islands such as emergent land features on atolls can be called islets, skerries, cays or keys.

Which of the two pictures is an Island?
Put a check mark on the circle.

DESERT

A **desert** is a barren area of land where little precipitation occurs and consequently living conditions are hostile for plant and animal life.

Which of the two pictures is a Desert? Put a check mark on the circle.

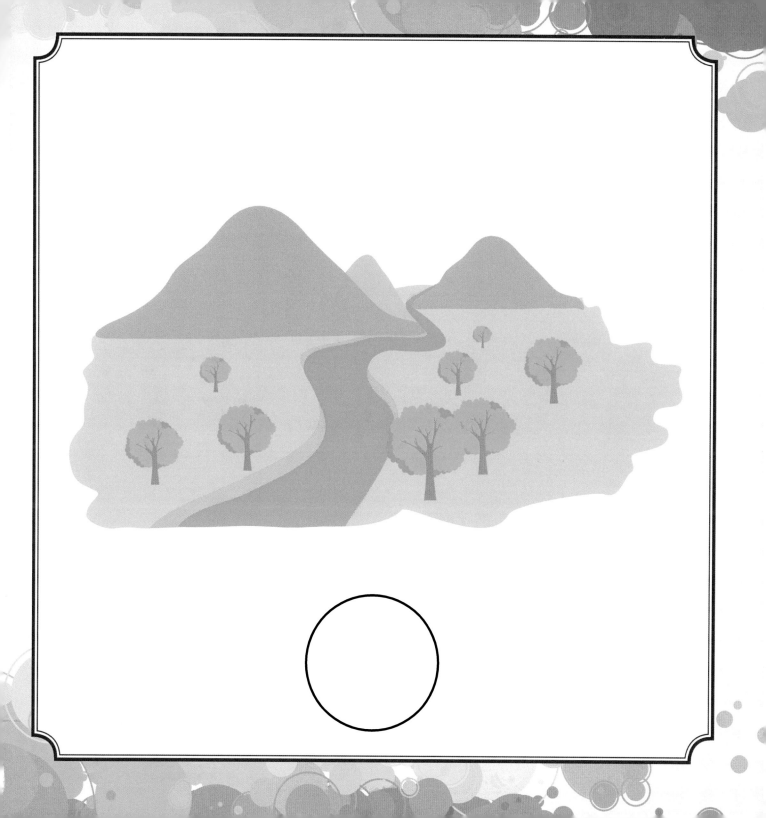

FOREST

A **forest** is a large area of land covered with trees or other woody vegetation. Forests are the dominant terrestrial ecosystem of Earth, and are distributed across the globe.

Which of the two pictures is a Forest Put a check mark on the circle.

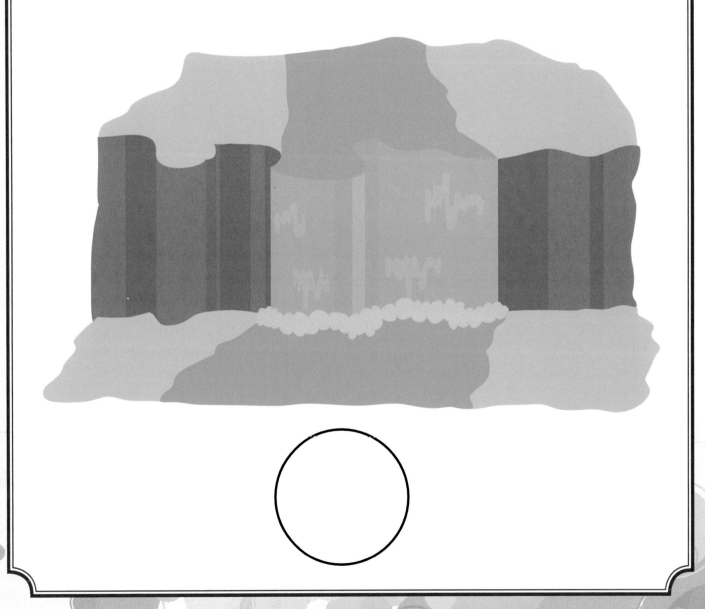

MOUNTAINS

A **mountain** is a large landform that stretches above the surrounding land in a limited area, usually in the form of a peak. A mountain is generally steeper than a hill. Mountains are formed through tectonic forces or volcanism.

Which of the two pictures is a Mountain? Put a check mark on the circle.

PLAIN

A **plain** is a flat area. Plains occur as lowlands and at the bottoms of valleys but also on plateaus or uplands at high elevations.

Which of the two pictures is a Plain? Put a check mark on the circle.

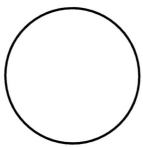

Made in the USA
Monee, IL
10 October 2020